Quick Reference to Food Safety & Sanitation

Nancy R. Rue, Ph.D.

Anna Graf Williams, Ph.D.

Upper Saddle River, NJ 07458

Editor-in-Chief: Steve Helba
Acquisitions Editor: Vernon Anthony
Managing Editor: Mary Carnis
Production Editor: Brian Hyland
Director of Manufacturing and Production: Bruce Johnson
Manufacturing Buyer: Ilene Sanford
Design Director: Cheryl Asherman
Senior Design Coordinator: Miguel Ortiz
Editorial Assistant: Ann Brunner
Desktop Publisher: Karen J. Hall
Development Editor: Anna Graf Williams
Printer/Binder: RR Donnelley - Crawfordsville
Cover Design: Amy Rosen
Cover Illustration: John Wise

Pearson Education LTD.
Pearson Education Australia PTY, Limited
Pearson Education Singapore, Pte. Ltd
Pearson Education North Asia Ltd
Pearson Education Canada, Ltd.
Pearson Educación de Mexico, S.A. de C.V.
Pearson Education -- Japan
Pearson Education Malaysia, Pte. Ltd

10 9 8 7 6 5 4 3 2 1
ISBN 0-13-042402-1

Contents

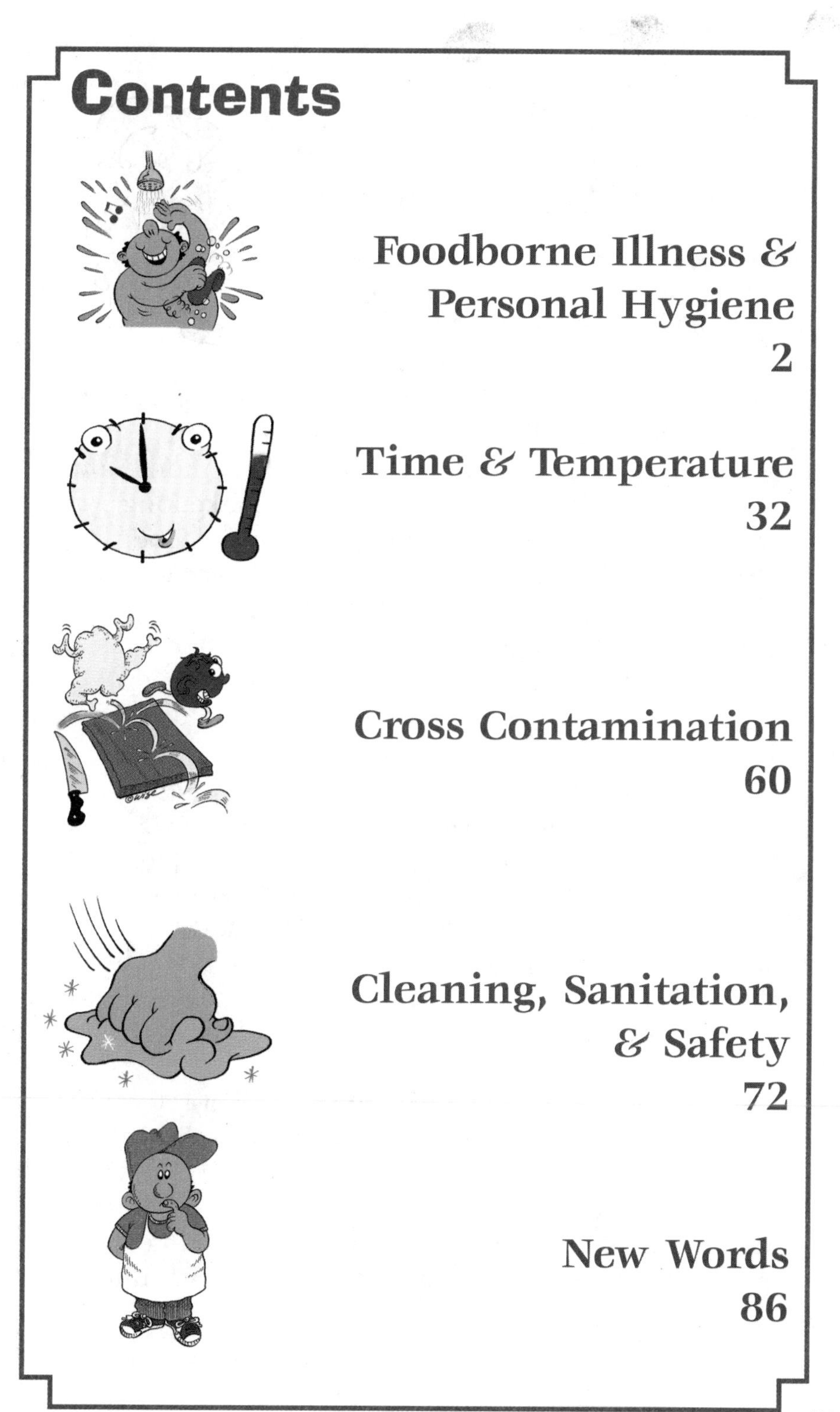

Preface

In food establishments, food workers are the front line of defense against the contamination of food and food products. The purpose of this book is to deliver critical food-handling information in a quick, easy-to-understand format. All the information has been conveyed in simple language and through descriptive pictures and illustrations to demonstrate the principles of food safety. It will not take long to master important procedures needed for food protection. After a very short course of instruction, the food worker should be able to apply this information to the performance of his or her tasks, and, ultimately, better protect the safety of food before it reaches the consumer.

Food Manager Certification is required in many states and jurisdictions across the United States. The manager is responsible for teaching employees how to keep food safe during receiving, storage, preparation, holding, and service. The Quick Reference is designed to assist managers in teaching employees the basics in food safety information and practice.

On the back cover, you will find the pocket reference guide for safe time and temperature controls. Cut it out for use as an easily accessible source to the critical information needed to keep food safe. With this book, you and your staff will be practicing proper food safety techniques in a very short time!

"We are gathered here today to talk about something important — Food Safety."

Bad things can happen when foodborne illness occurs:

- Customers may not want to eat at the establishment anymore
- The establishment could go out of business
- You could be fired
- Someone could die.

Section 1

FOODBORNE ILLNESS & PERSONAL HYGIENE

Where do *you* want to eat today?

Foodborne illness happens when people eat contaminated food and get sick. It can be stopped from ever happening. The #1 thing that can stop foodborne illness is YOU! In the next several pages, you will find out how YOU can stop foodborne illness and why YOU are so important in keeping food safe for people to eat. The first section of this book will tell you what you should do before you come to work and what you should do while you are working to help keep food safe. Pay close attention; there will be a test at the end of each section!

NEW WORDS:

Bacteria
Biological Hazards
Contaminated
Germs
Parasites
Personal Hygiene
Virus

"What is foodborne illness?"

Foodborne illness is when people get sick from eating food that contains germs, chemicals, or other harmful materials.

Germs:
tiny organisms that are too small to be seen by the naked eye and can cause illness

Have you ever had a foodborne illness? Many of us have, but did not even realize it! People who get sick with diarrhea, fever, vomiting, and other symptoms may not think "bad food" is the cause of their illness, but it could be.

Contaminated:
the presence of harmful germs, chemicals, or non-food items

"Did you know there are very specific causes of foodborne illness?"

Biological Hazards

Chemicals

-and-

Physical hazards like pieces of glass

Biological Hazard: bacteria, viruses, and parasites in food that make people sick

"Some possible causes of foodborne illness are":

Bacteria

Bacteria:
germs, some of which can make you sick
Example: *Salmonella spp.*

Viruses

Virus:
a germ that lives on or in other animals and humans
Example: Hepatitis A virus

Parasites.

Parasites:
plants or animals that live and feed in or on another plant or animal
Example: *Trichinella*

More things that can cause foodborne illness or injury ...

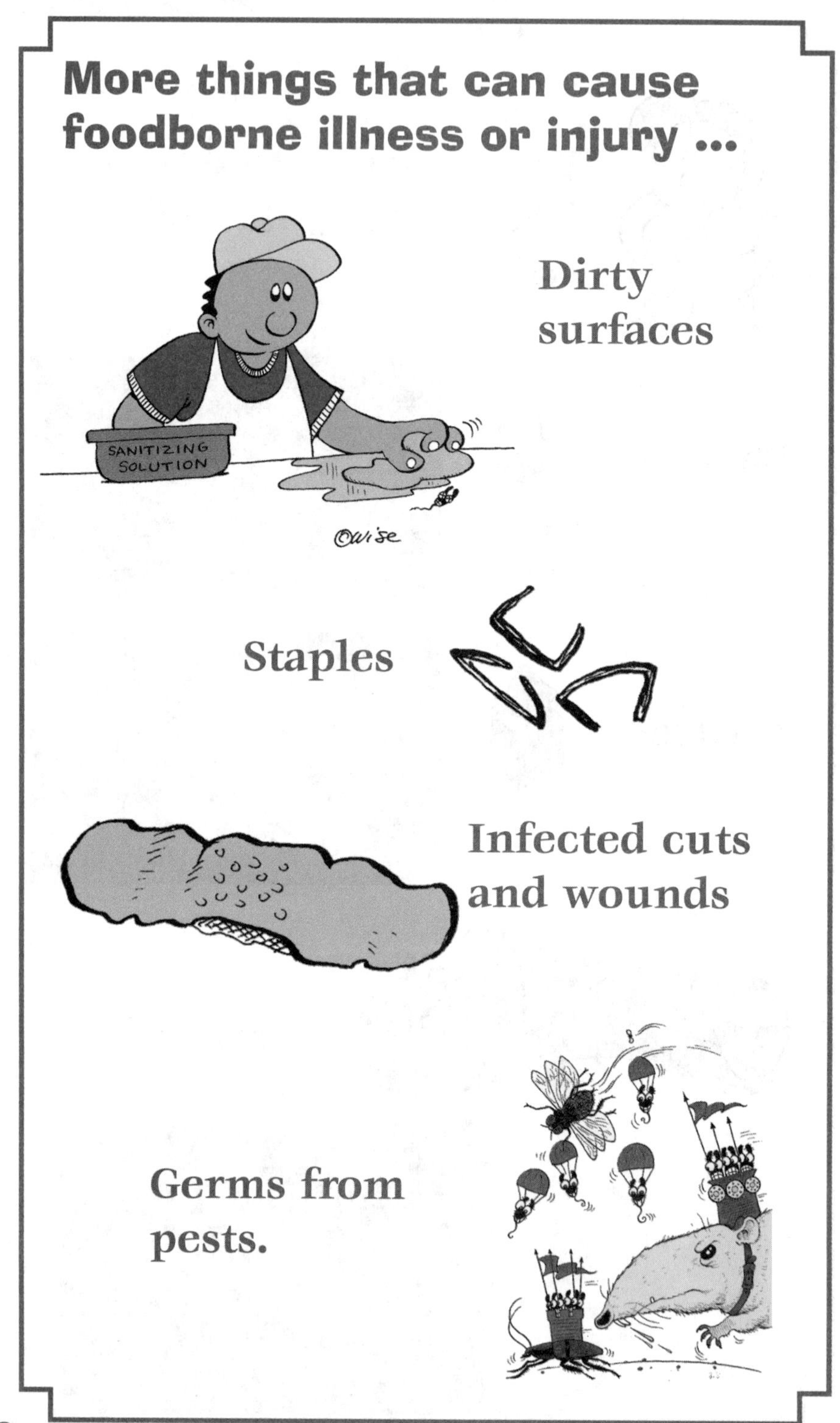

Dirty surfaces

Staples

Infected cuts and wounds

Germs from pests.

"You should not work with food if ..."

You have open cuts or wounds, unless you cover them with bandages and gloves.

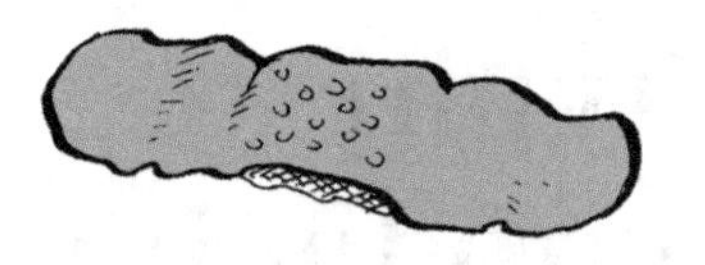

You are sick with nausea, vomiting, or diarrhea.

"If you have been told you currently have or have had the following in the past three months:

- Hepatitis A virus
- *Shigella spp.*
- *Salmonella* Typhi
- Shiga toxin-producing *Escherichia coli,*

you must tell your boss. Also tell your boss if you think you may have been near others with these diseases."

Your daily personal hygiene can also affect foodborne illness.

Personal Hygiene: health habits including bathing, washing hair, wearing clean clothing, and proper hand washing

"Good hygiene includes":

1. Showering every day before work and using deodorant or antiperspirant
2. Keeping your hair neat and clean
3. Keeping your fingernails short and clean
4. Wearing clean clothes and uniforms
5. Wearing no jewelry except a plain wedding band
6. Not using tobacco in any form around food preparation areas, equipment, and sink areas
7. Covering cuts and wounds on fingers and hands with water-resistant bandages and a single-use glove
8. Putting your personal items in areas away from food or where food is prepared and stored
9. Washing your hands properly and frequently.

Always ...

take a shower before going to work.

wear clean clothes and closed-toed shoes.

WEAR A CLEAN APRON EVERY DAY

change your apron when it gets dirty!

"An important part of personal hygiene is washing your hands."

Wash your hands:

when you arrive at
your work center.

Wash your hands:

after using
the toilet.

Wash your hands:
after
eating or drinking,
smoking,
coughing or sneezing,
touching your body,
or using tissues or handkerchiefs.

Wash your hands:

between working with raw foods and ready-to-eat foods.

Wash your hands:

after handling dirty
or soiled equipment.

Wash your hands:

after taking out garbage.

after wiping counters.

after mopping.

after handling money.

Wash your hands:

after changing a baby's diaper.

Wash your hands:

**after you touch
a live animal.**

"There is a correct place to wash your hands ..."

Always wash your hands in a hand sink that has:

- Hot and cold running water
- Soap
- Paper towels or an air dryer.

Do not wash your hands in food preparation or warewashing sinks.

"Did you know there is a right way to wash your hands?"
SOAP
©wise

"Here are the steps to washing your hands ..."
SOAP
1. Wet Hands
2. Apply Soap
3. Briskly Rub Hands for Twenty Seconds
4. Scrub Fingertips and Between Fingers
5. Scrub Forearm to Just Below Elbow
6. Rinse Forearms and Hands
TOWELS
7. Dry Hands and Forearms
8. Turn Off Water Using a Paper Towel
9. Turn Doorknob and Open Door Using Paper Towel
10. Discard Towel

Gloves are used to protect your customers, not to protect your hands from getting dirty.

You should also change your gloves after you ...

handle money,

touch dirty things,

touch your hair or any other part of your body,

change from handling raw food (like chicken) to ready-to-eat foods (like fresh vegetables).

Never re-use a pair of gloves. Always throw away used gloves and get a new pair.

**"Remember ...
everything you do
makes food safe or
unsafe."**

Wash your
hands.

Use gloves the
right way.

Don't smoke in
your work area.

Don't eat or drink in work areas.

Change your uniform or apron when it gets dirty.

Keep personal items out of food production areas.

True or False

T F 1. You should always wash your hands after going to the toilet.

T F 2. If you have a cut or wound and cover it with a bandage and a single-use glove, you can work with food.

T F 3. Foodborne illness cannot be prevented.

T F 4. You should always wash your hands in food preparation and warewashing sinks.

1. T, 2. T, 3. F, 4. F

Section 2

TIME & TEMPERATURE

Food is handled many times from the moment it comes into the place you work until the time someone eats it. It is your job to make sure food stays safe to eat. One important thing you have to do is make sure food stays at the right temperatures for the right amount of time. This section will tell you how to take the temperatures of food and at what temperatures different foods should be kept while being handled.

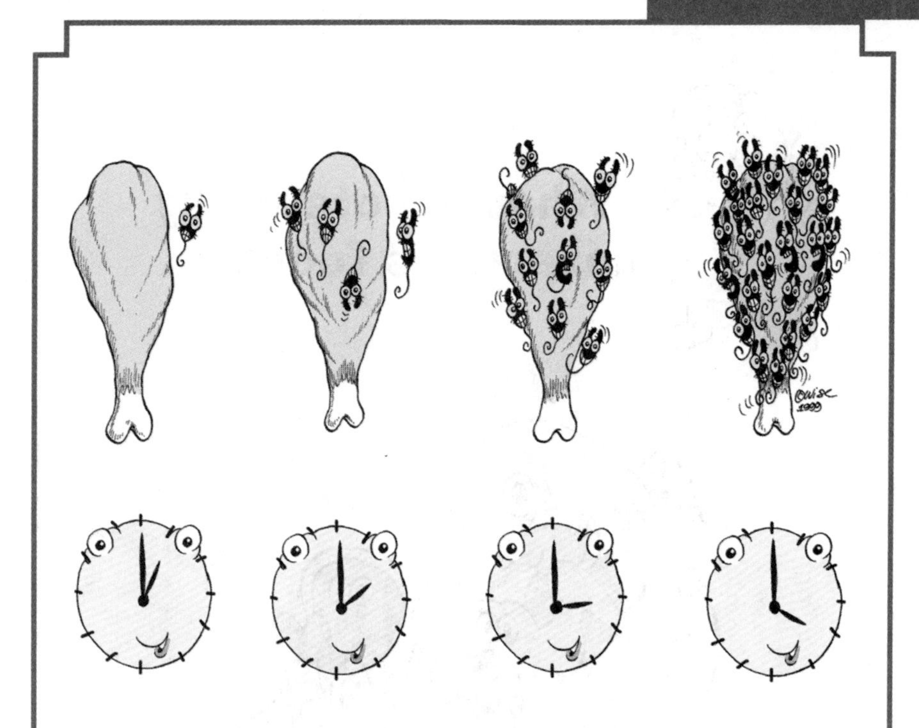

NEW WORDS:

Potentially Hazardous Food
Food Temperature Danger Zone
Flow of Food
FIFO

"What is potentially hazardous food?"

Potentially Hazardous Food:
food capable of supporting the rapid and progressive growth of harmful germs

"Where are germs found?"
"They are everywhere!!!"

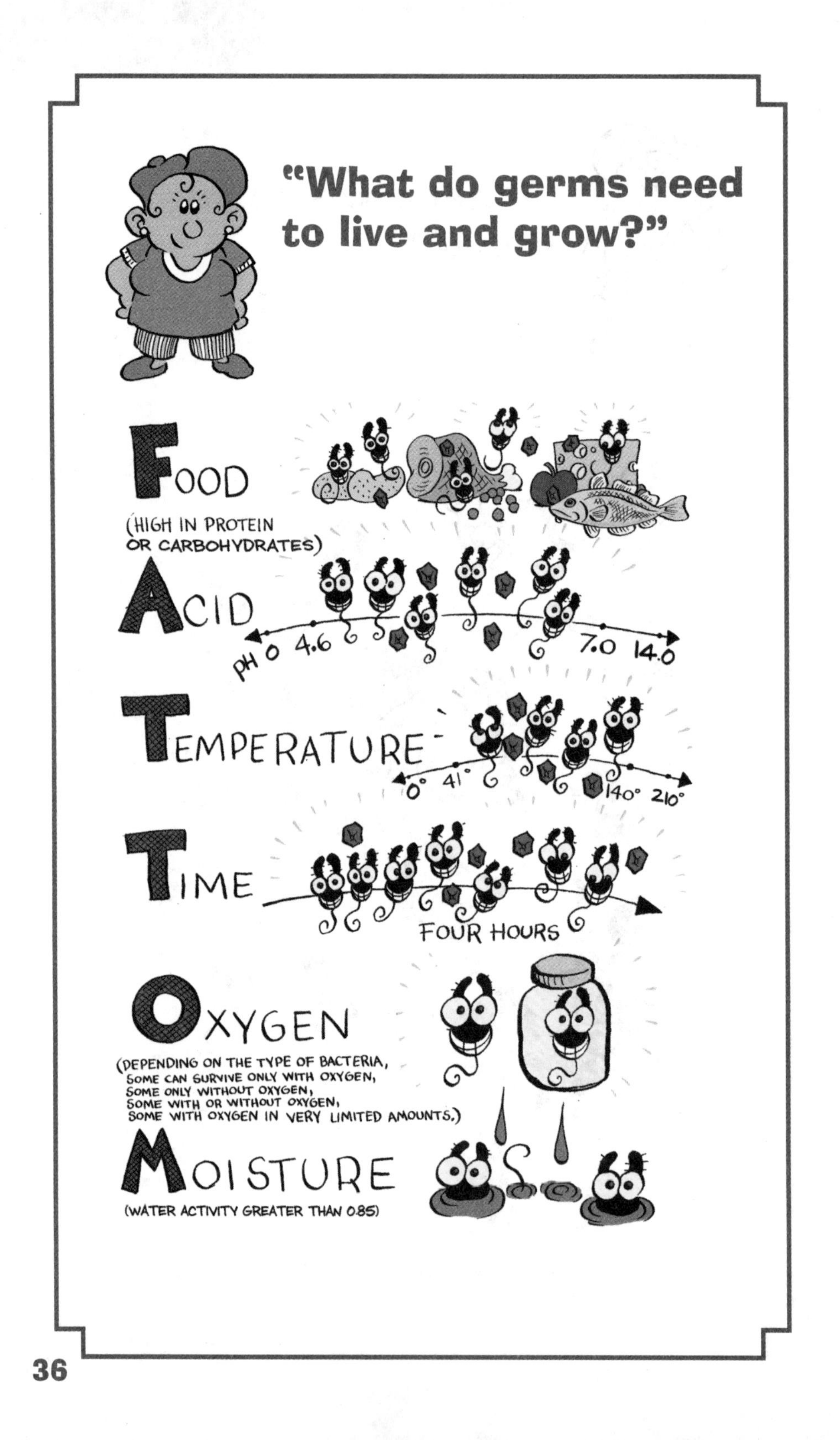

"What do germs need to live and grow?"
FOOD
(HIGH IN PROTEIN OR CARBOHYDRATES)
ACID
pH 0 4.6 7.0 14.0
TEMPERATURE
0° 41° 140° 210°
TIME
FOUR HOURS
OXYGEN
(DEPENDING ON THE TYPE OF BACTERIA,
SOME CAN SURVIVE ONLY WITH OXYGEN,
SOME ONLY WITHOUT OXYGEN,
SOME WITH OR WITHOUT OXYGEN,
SOME WITH OXYGEN IN VERY LIMITED AMOUNTS.)
MOISTURE
(WATER ACTIVITY GREATER THAN 0.85)

Foods held between 41°F (5°C) and 140°F (60°C) are considered to be in the Food Temperature Danger Zone (TDZ).

Food Temperature Danger Zone: temperatures between 41°F (5°C) and 140°F (60°C) at which bacteria grow best

The TDZ is where foodborne illness-causing bacteria grow best.

Food can only be held in the TDZ for a maximum of four hours.

Time & Temperature

"When dealing with potentially hazardous food ... keep it hot at 140° F (60°C) or above, keep it cold at 41°F (5°C) or below, or don't keep it at all."

"Your goal is to keep food OUT of the food temperature danger zone."

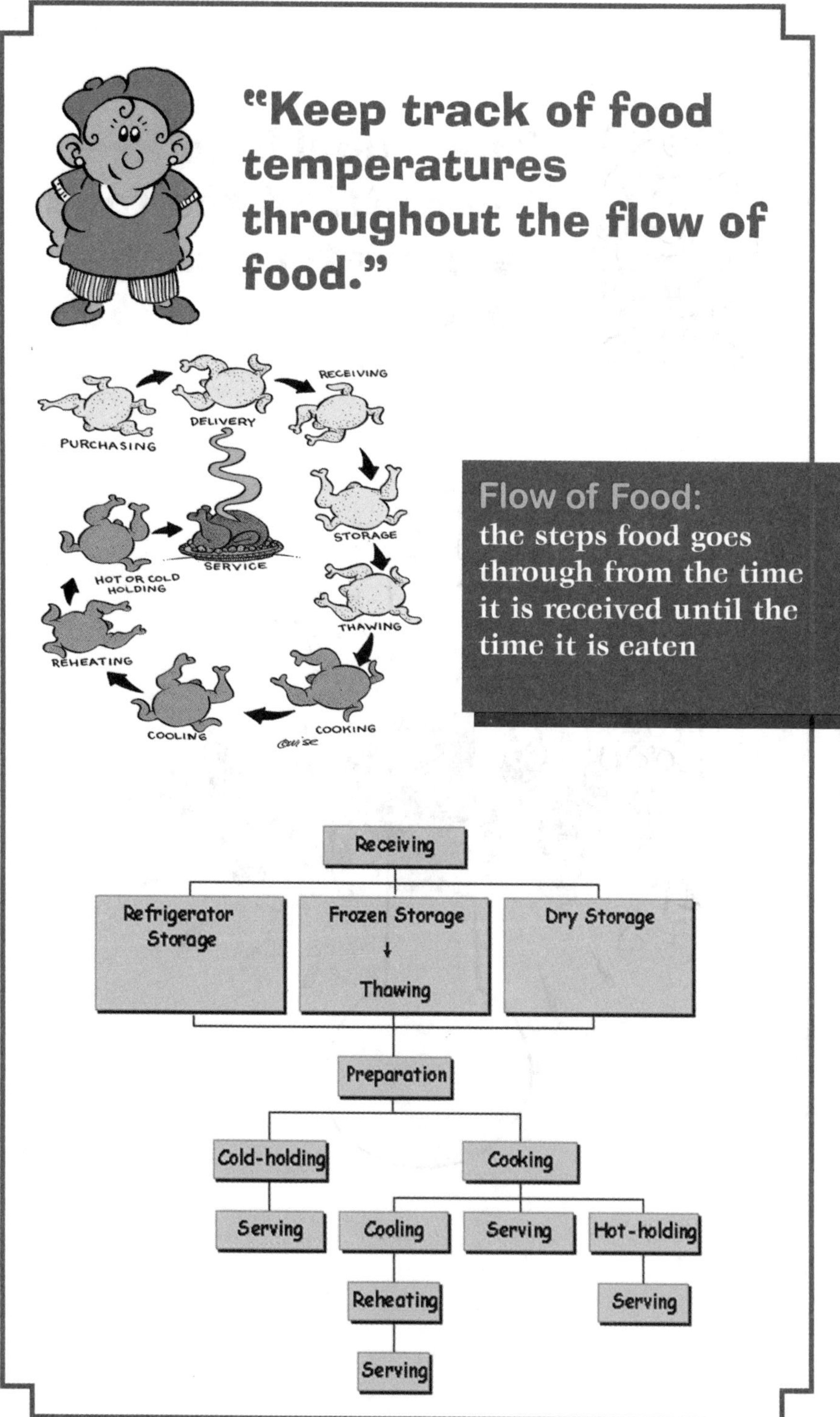
"Keep track of food temperatures throughout the flow of food."
PURCHASING
DELIVERY
RECEIVING
STORAGE
THAWING
COOKING
COOLING
REHEATING
HOT OR COLD HOLDING
SERVICE
Flow of Food:
the steps food goes through from the time it is received until the time it is eaten
Receiving
Refrigerator Storage
Frozen Storage
Thawing
Dry Storage
Preparation
Cold-holding
Cooking
Serving
Cooling
Serving
Hot-holding
Reheating
Serving
Serving

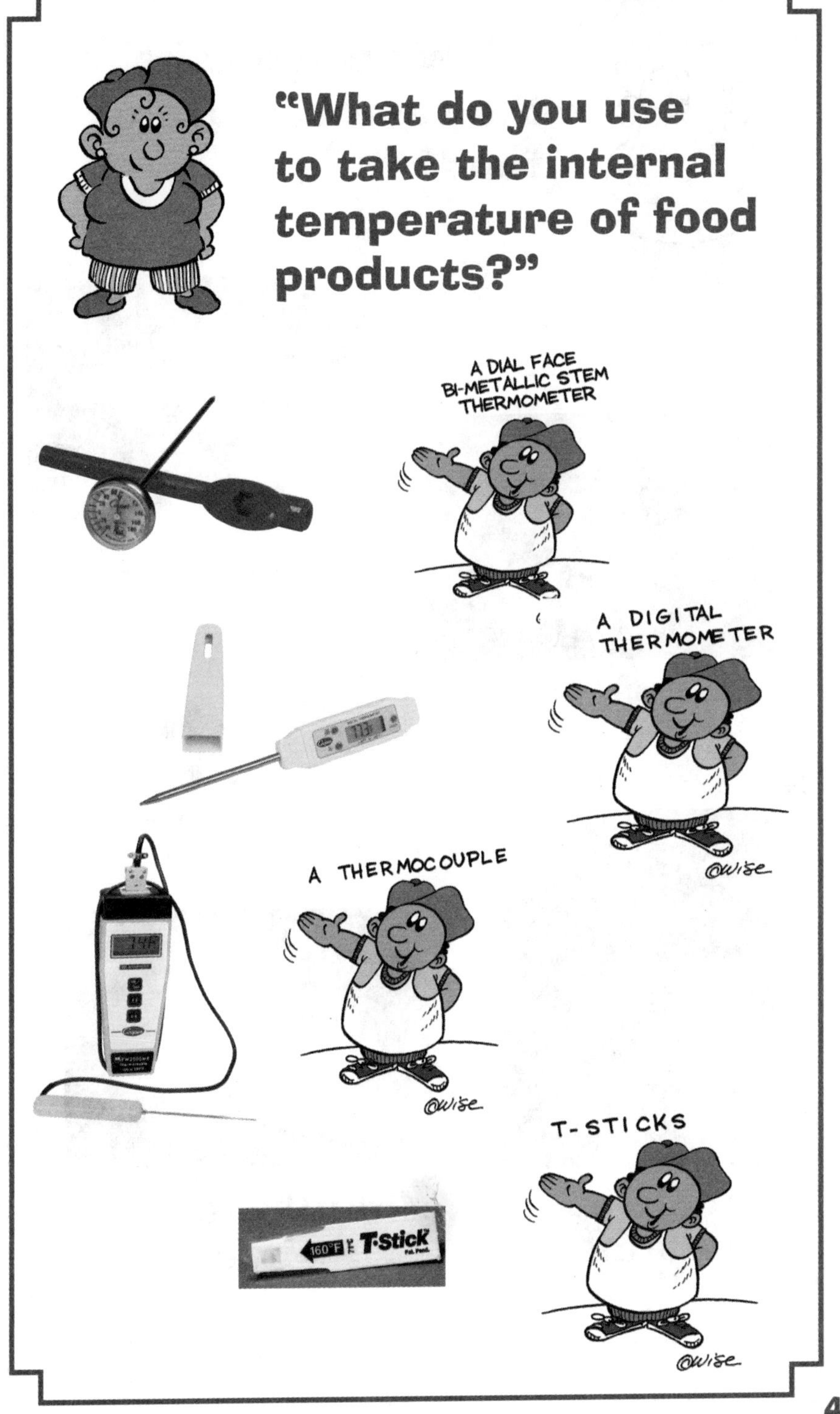
"What do you use to take the internal temperature of food products?"
A DIAL FACE BI-METALLIC STEM THERMOMETER
A DIGITAL THERMOMETER
©Wise
A THERMOCOUPLE
©Wise
T-STICKS
160°F
T-Stick
©Wise

"How do you take the temperature of food?"

Salad bar

Prepackaged meats

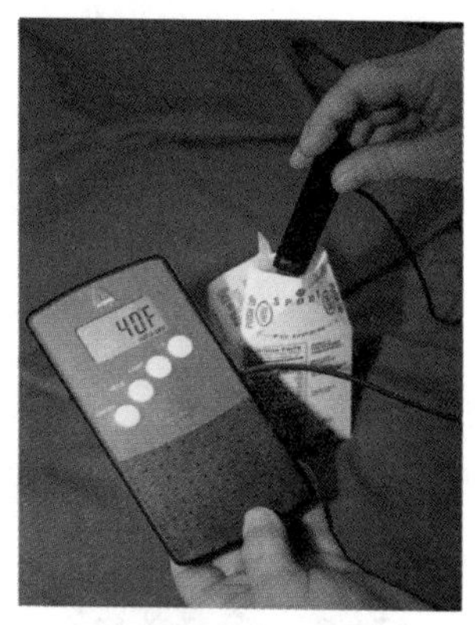

Single-use containers

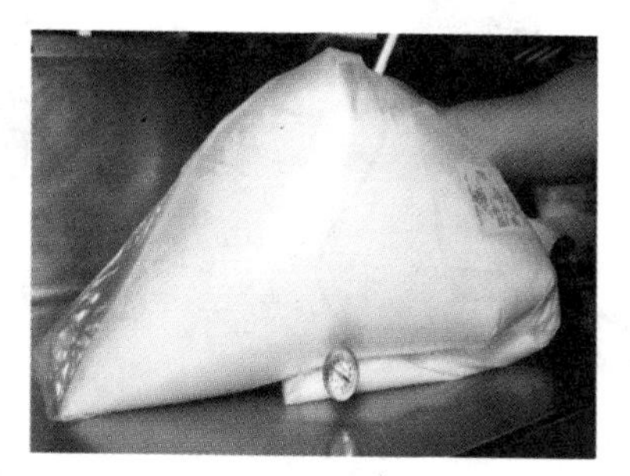

Bulk storage items

"This is how you use a thermometer ..."

Insert the probe of the thermometer 2 inches into the food product or between packaged foods and wait until the needle or numbers stop.

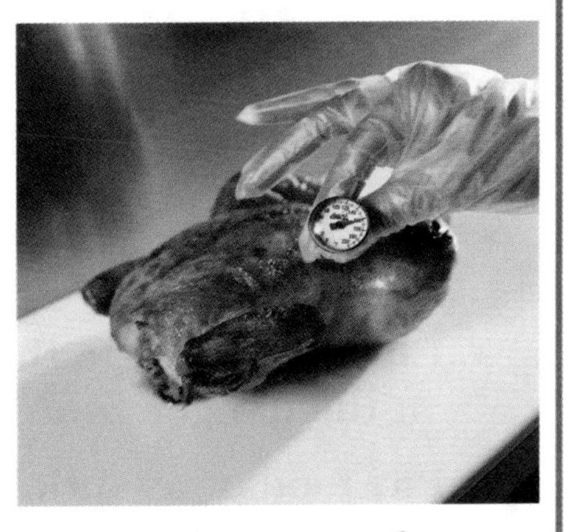

Be sure to clean and sanitize the thermometer before and after each use!

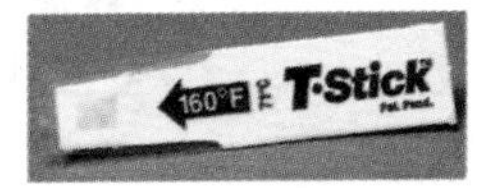

T-Sticks change color at the temperature they measure.

"When food products are received, you must receive and store them in the right way."

1. Check to be sure all product is there
2. Check the product's packaging and do not take any damaged product
3. Check to be sure the temperature is not in the danger zone
4. Put food away quickly and correctly

Be sure to store cooked or ready-to-eat items above raw foods and store everything at least 6 inches above the floor.

"When putting away products, use the FIFO inventory method."

FIFO:
First In - First Out

put new products behind old products

Thawing Food

"When you thaw food, move it through the food temperature danger zone quickly."

Thaw foods in a microwave.

Foods thawed in a microwave oven must be cooked to completion in the microwave or transferred immediately to a stove, oven, or other type of equipment to complete the cooking process.

Thaw foods as part of the cooking process.

Thaw food submerged under cool running water.

- Ensure food is completely submerged under cool running water [70 °F (21 °C) or below].
- Ready-to-eat foods must not reach the temperature danger zone at all, and raw animal foods must not be in the temperature danger zone for more than four hours.
- Thawing counts toward time in the temperature danger zone.

The most preferred method is to thaw foods in the refrigerator.

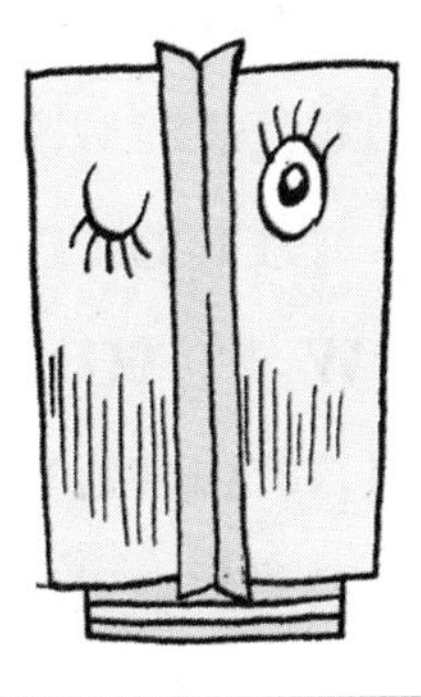

Cooking Food

When you are preparing or cooking food, it is your job to keep it safe!

Food can become contaminated with germs during the cooking and preparation process if you allow it to remain in the food temperature danger zone for more than four hours.

"There are several things you need to keep in mind when trying to keep food safe."
Temperature
Time
SAFE FOOD
Cold
Heat
©Wise
Wash-Rinse-Sanitize
Hand Washing

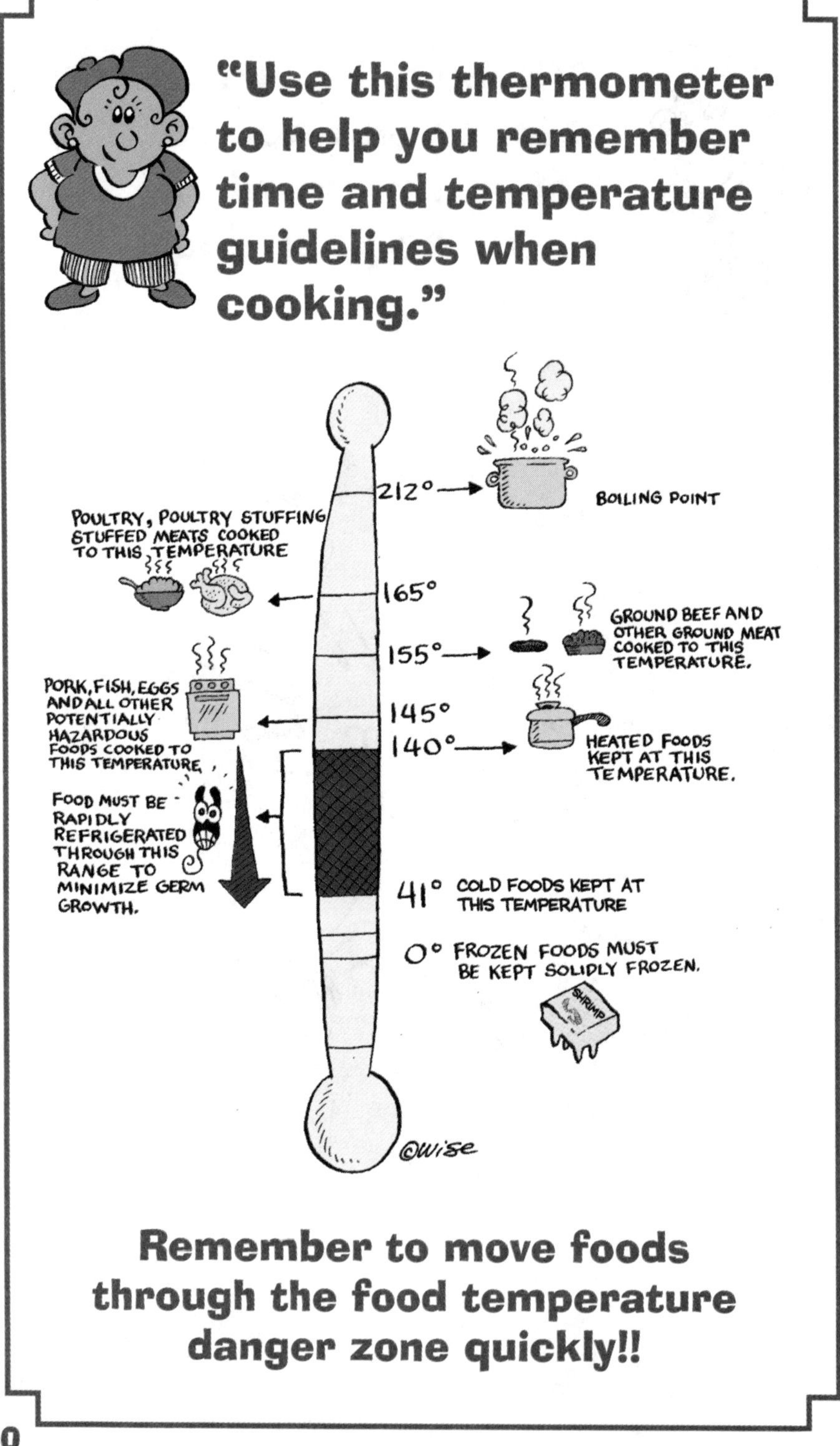
"Use this thermometer to help you remember time and temperature guidelines when cooking."
212°
BOILING POINT
POULTRY, POULTRY STUFFING STUFFED MEATS COOKED TO THIS TEMPERATURE
165°
155°
GROUND BEEF AND OTHER GROUND MEAT COOKED TO THIS TEMPERATURE.
PORK, FISH, EGGS AND ALL OTHER POTENTIALLY HAZARDOUS FOODS COOKED TO THIS TEMPERATURE
145°
140°
HEATED FOODS KEPT AT THIS TEMPERATURE.
FOOD MUST BE RAPIDLY REFRIGERATED THROUGH THIS RANGE TO MINIMIZE GERM GROWTH.
41°
COLD FOODS KEPT AT THIS TEMPERATURE
0°
FROZEN FOODS MUST BE KEPT SOLIDLY FROZEN.
SHRIMP
©Wise
Remember to move foods through the food temperature danger zone quickly!!

Hot & Cold Food Holding

"Now that the food is cooked, we have to think about how to keep harmful germs from growing before the food is eaten."

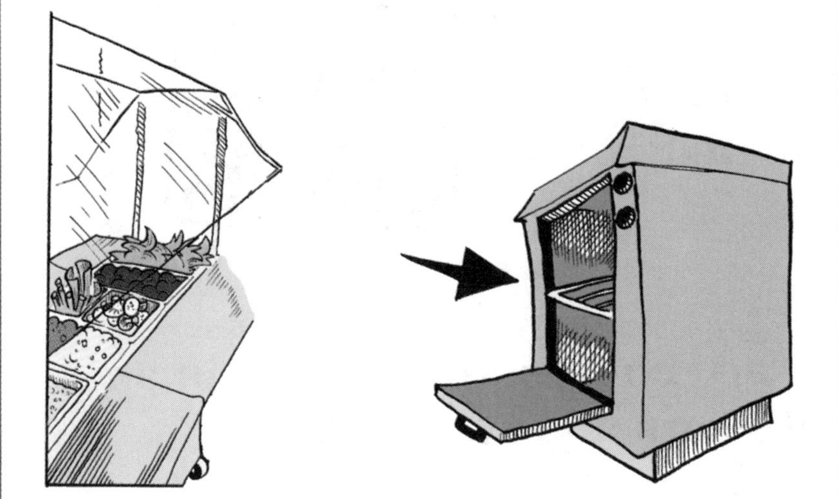

Keep food at temperatures below 41° (5°C) or above 140°F (60°C).

Cooling Food

"Good ways to cool food are":

Add ice as a final ingredient

Place hot food containers in an ice bath and stir

Stir food frequently to speed up cooling

Separate larger portions into smaller portions to help it cool faster.

Cool foods rapidly from 140°F (60°C) to 70°F (21°C) within the first two hours AND from 140°F (60°C) to 41°F (5°C) or below within six hours.

Cold Storage

Label, date, and cover all food.

Vent foods that are cooling until they reach 41 °F (5 °C) or below.

Leave space between containers to allow air to circulate.

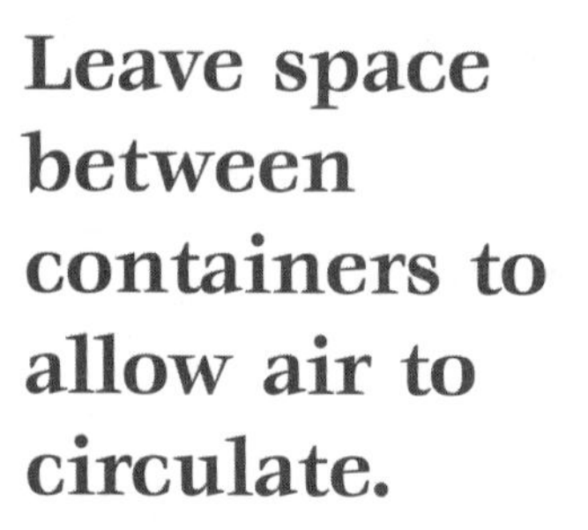

Store raw products below ready-to-eat products.

Reheating for Hot Holding

Hold food at 41°F (5°C) or below until you are ready to reheat it.

Reheat food to 165°F (74°C) within two hours.

Reheat food as few times as possible!

"Pull out this handy time and temperature chart and use it to help you remember food safety rules."

Time and Temperature

Receiving and Storing:

Frozen and refrigerated receiving/storage practices prevent or slow the growth of harmful microorganisms

Food Product	Internal Temperature	Times
Frozen Foods	Solidly frozen 0°F (-18°C) recommended	Weeks & Months
Refrigerated Foods	41°F (5°C) or lower	As food quality allows
Raw Shell Eggs	45° F (7° C) or below ambient temperature	Until sell-by date has expired

Thawing:

Take food from frozen to non-frozen to minimize the product's time in the temperature danger zone. Keep ready-to-eat foods below 41°F (5°C) at all times.

Method	Internal Temperature	Times
In refrigerator	41°F (5°C) or lower	typically takes 2-3 days
Submerged under cool running water 70°F (21°C)	41°F (5°C) or lower	4 hours or less (counts toward time in the food temperature danger zone).

Cooking:

Safely getting a food product from raw to ready-to-eat with minimum time held at internal temperature before serving

Food Product	Minimum Internal Temperature	Times
Beef Roast (rare)	130°F (54°C) 140°F (60°C)	112 minutes 12 minutes
Beef (other than roasts), Pork (other than roasts), Fish	145°F (63°C)	15 seconds
Ground Beef, Ground Pork, Game Animals	155°F (68°C)	15 seconds
Beef Roast (medium), Pork Roast, Ham	145°F (63°C)	4 minutes
Poultry, Stuffed Meats, Stuffed Food Products	165°F (74°C)	15 seconds

2001 Food Code

Time and Temperature (continued)

Hot-Holding:

Keeping hot food out of the temperature danger zone

Food Product	Internal Temperature	Times
Hot-holding of all foods	**140°F (60°C) or above**	**Until product quality is unacceptable**

Cold Food Holding:

Keeping cold food out of the temperature danger zone

Food Product	Internal Temperature	Times
Cold-holding of all foods	**41°F (5°C) or below**	**Until product quality is unacceptable or sell-by date has expired**

Cooling Hot Foods:

Rapid reduction of temperature through and out of the temperature danger zone

Part	Internal Temperature	Times
Hot Food Cooling part 1	**From 140° to 70°F (60° to 21°C)**	**2 hours or less**
Hot Food Cooling part 2	**From 140° to 41°F (60° to 5°C) or below**	**Within 6 hours or less**

Frozen Food Holding: **Keeping food solidly frozen**

Food Product	Internal Temperature	Times
Frozen Food	**Solidly Frozen 0°F (-18°C) recommended**	**Until product quality is unacceptable**

Reheating: **Bringing food back up to serving temperature**

Method	Internal Temperature	Times
Reheating	**165°F (74°C) or above**	**Within 2 hours**

Remember, there's NEVER been a case of foodborne illness that couldn't have been prevented!

2001 Food Code

"Food safety doesn't end there ... You still have to bring it to the customer."
DO
DON'T
Handle utensils properly.
DO
DON'T
WRONG
Use proper carrying procedures.
RIGHT
"Now that's good service!"

True or False

T F 1. The Food Temperature Danger Zone is between 41°F (5°C) and 140°F (60°C).

T F 2. If you do not know how long food has been out at room temperature, you should refrigerate it as soon as possible.

T F 3. Germs can grow when held at temperatures between 41°F (5°C) and 140°F (60°C).

T F 4. It is not important to track food temperatures throughout the flow of food.

1. T, 2. F, 3. T, 4. F

Section 3

CROSS CONTAMINATION

Everything that touches food can make it unsafe to eat. It is very important you keep food from touching or coming in contact with other foods or things that can contaminate them. Did you know there are tiny creatures you cannot even see that can make food unsafe to eat? This section will tell you what you should keep away from food and how you can keep those germs from growing on food.

NEW WORDS:

Cross Contamination
Sanitize

"Help Me...
stay away from
Cross Contamination!"
Cross
Contamination:
transfer of harmful
germs between
items

"How Can You Help?"
"First, let me show you what cross contamination is."

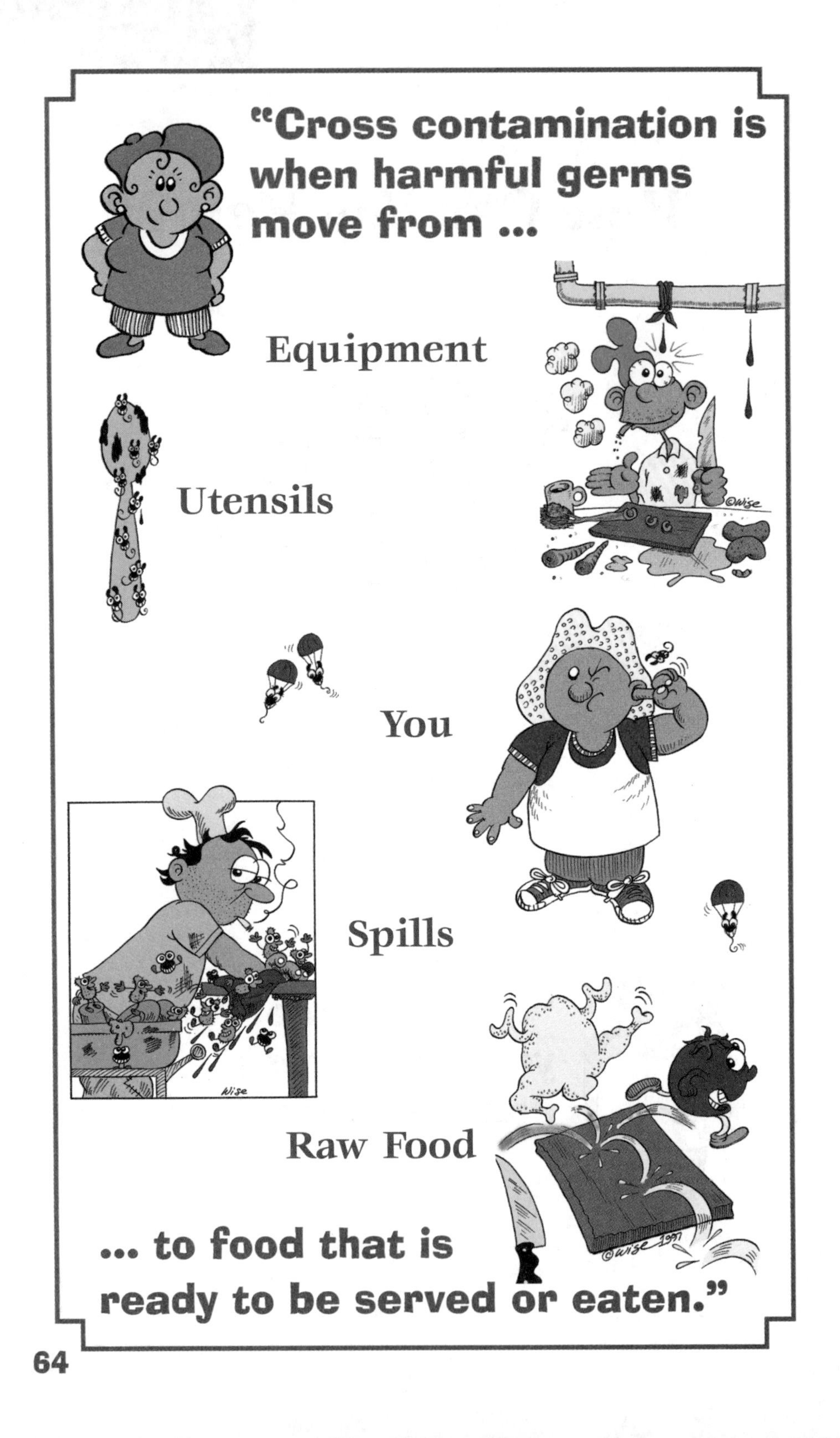
"Cross contamination is when harmful germs move from ...
Equipment
Utensils
You
Spills
Raw Food
... to food that is ready to be served or eaten."
©Wise
Wise
©Wise 1997

"You cannot see, smell, or taste the harmful germs that cause foodborne illness!"
POTATOES
MILK
POTATOES
MILK
DIP

"The harmful germs that you cannot see, smell, or taste are called biological hazards."

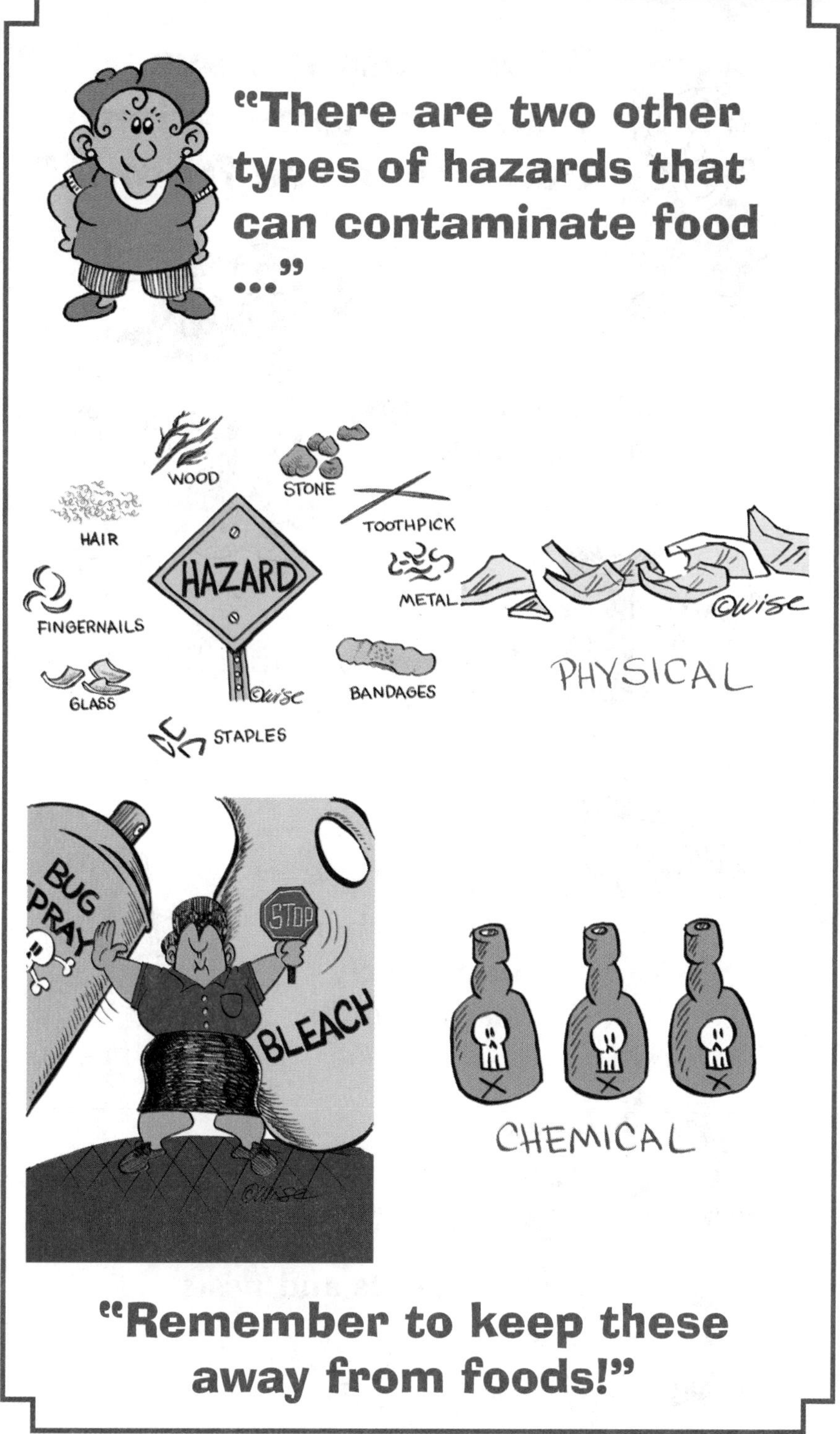
"There are two other types of hazards that can contaminate food ..."
WOOD
STONE
HAIR
TOOTHPICK
HAZARD
METAL
FINGERNAILS
GLASS
BANDAGES
STAPLES
PHYSICAL
BUG SPRAY
STOP
BLEACH
CHEMICAL
"Remember to keep these away from foods!"

"Some other ways cross contamination can occur are from ..."

Chemical residues from fruits and vegetables

Workers touching live animals

The worker's mouth when improperly tasting foods

Rodents and pests.

"There are several ways you can help stop cross contamination ..."

Wash your hands often

Store raw foods below and away from cooked foods. Store cooked and ready-to-eat foods on the top shelf, then raw beef on the next shelf down, raw fish and shellfish on the next shelf down, and raw poultry on the bottom shelf

Wear single-use gloves or use proper utensils.

"You should always clean and sanitize utensils and equipment when changing from raw to cooked or ready-to-eat foods...

Sanitize:

reducing the number of harmful germs to a safe level

... or from one species of food to another, like cutting raw chicken and then fish."

True or False

T F 1. Washing your hands with soap and water will help prevent cross contamination.

T F 2. It is OK to cut raw chicken on a cutting board and then chop lettuce on the same board without cleaning it.

T F 3. You cannot see germs with the naked eye.

T F 4. Three types of hazards are biological, chemical and physical.

1. T, 2. F, 3. T, 4. T

Section 4

Keeping your work area clean is very important. Did you know that just because something looks clean does not mean it is clean? Everything food touches must be cleaned as well as sanitized. Keep reading to find out how to clean and sanitize, and why it is so important to food safety!

NEW WORDS:

Clean
Sanitary

Cleaning & Sanitizing
Cleaning Steps
SOILED
CLEAN
SANITIZED
©wise 1999
Clean:
you cannot see dirt or food pieces
Sanitary:
the number of harmful germs has been reduced to a safe level
SANITIZED
©wise 1999

First clean ...
CLEANING SOLUTION
then rinse ...
then sanitize ...
SANITIZING SOLUTION
©Wise
... and let air dry.
Wise
You have now removed the harmful germs!

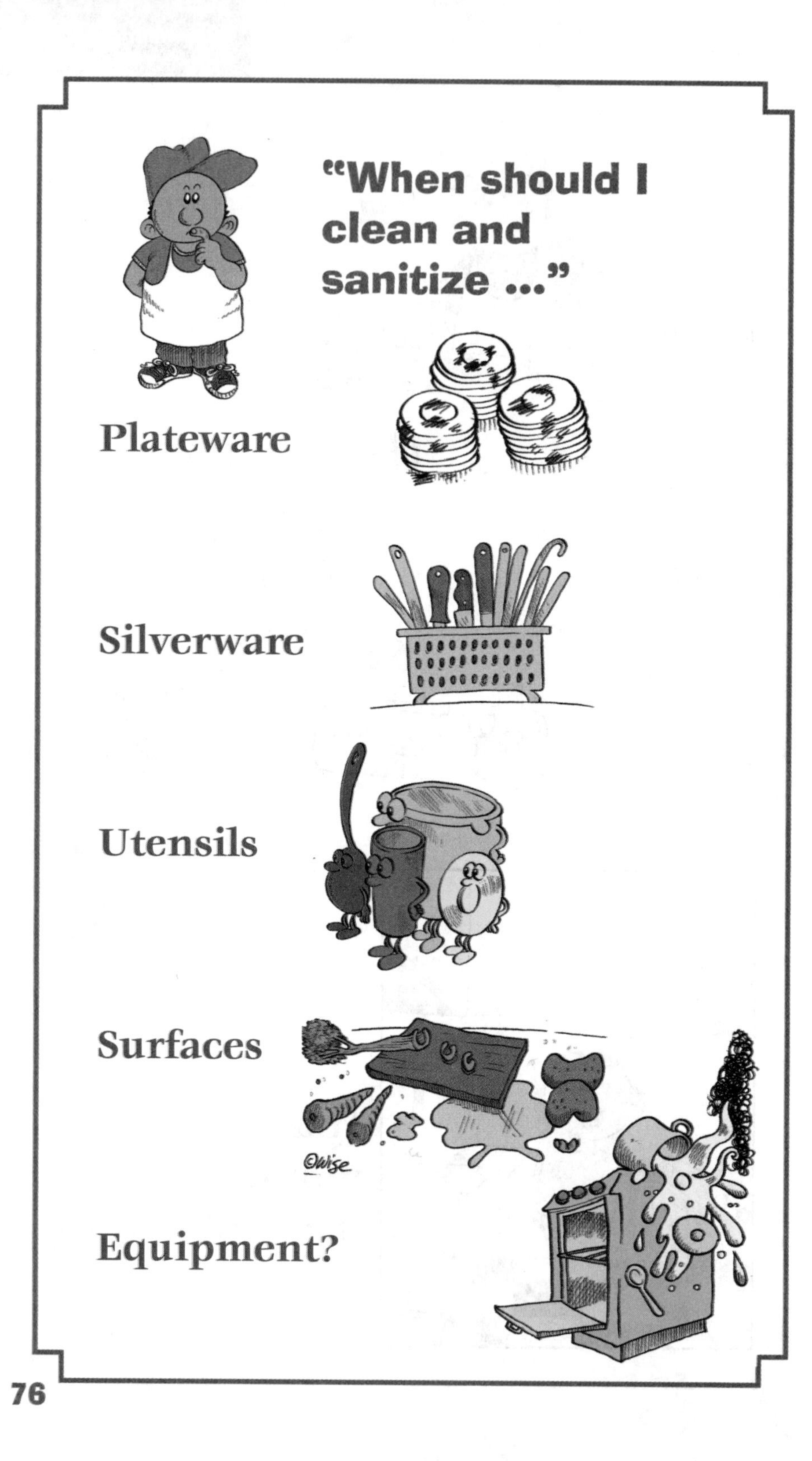
"When should I clean and sanitize ..."
Plateware
Silverware
Utensils
Surfaces
Equipment?
©Wise

"When they are dirty!"

"How do I clean and sanitize?"

Manually ...

Step 1. Wash in hot water with cleaner

Step 2. Rinse in clean water

Step 3. Sanitize in a warm chemical solution 75°F (24°C) to 120°F (49°C) or sanitize with 171°F (77°C) water

Step 4. Air dry.

or Mechanically ...

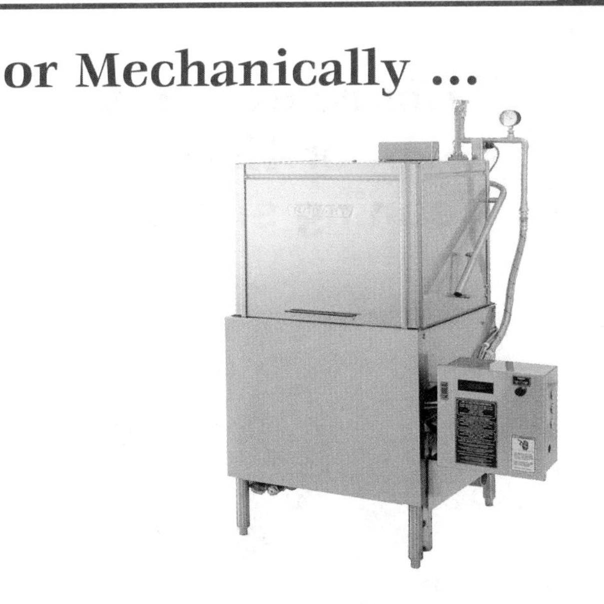

Step 1. Pre rinse

Step 2. Place objects to be cleaned in rack

Step 3. Place loaded rack into machine
The machine will:
a) Wash
b) Rinse
c) Sanitize

Step 4. Remove rack and allow objects to air dry.

"Where do I put, keep, or find cleaning supplies?"

Keep cleaning supplies in a secured area away from food.

"Be sure not to store chemicals next to foods."

"Protect yourself and your job! Follow all safety rules to prevent accidents."

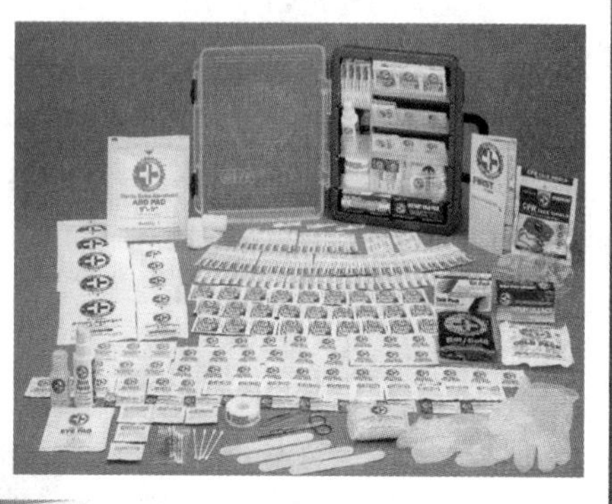

Know where the first aid kit is located.

Use caution signs when mopping floors.

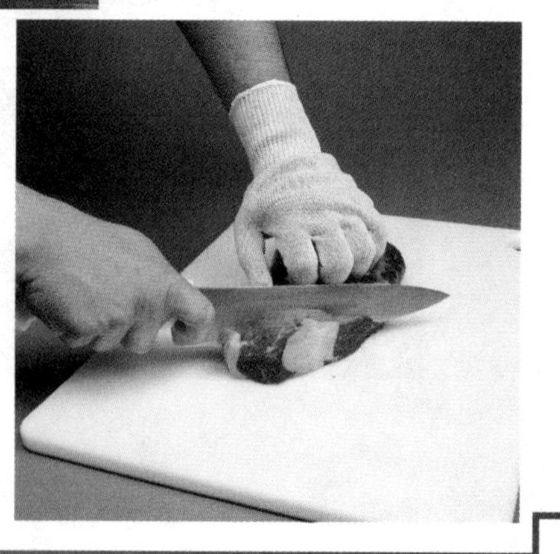

Use special gloves when cutting foods.

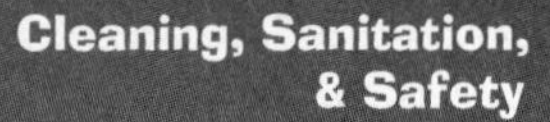

Know the locations of exits.

Know where fire extingushers are located and how to use them.

Turn power off when maintaining equipment.

"What kind of food worker do YOU want to be?"

True or False

T F 1. Cleaning is the same thing as sanitizing.

T F 2. Sanitizing should be done after cleaning.

T F 3. Chemicals may be stored next to food.

T F 4. All food contact surfaces must be air dried.

1. F, 2. T, 3. F, 4. T

NEW WORDS

Section 1
Foodborne Illness & Personal Hygiene

Bacteria:
germs, some of which can make you sick
Example: *Salmonella spp.*

Contaminated:
the presence of harmful germs, chemicals, or non-food items

Biological Hazard:
bacteria, viruses, and parasites in food that make people sick

Germs:
tiny organisms that are too small to be seen by the naked eye and can cause illness

Parasites:
plants or animals that live and feed in or on another plant or animal
Example: *Trichinella*

Personal Hygiene:
health habits including bathing, washing hair, wearing clean clothing, and proper hand washing

Virus:
a germ that lives on or in other animals and humans
Example: Hepatitis A virus

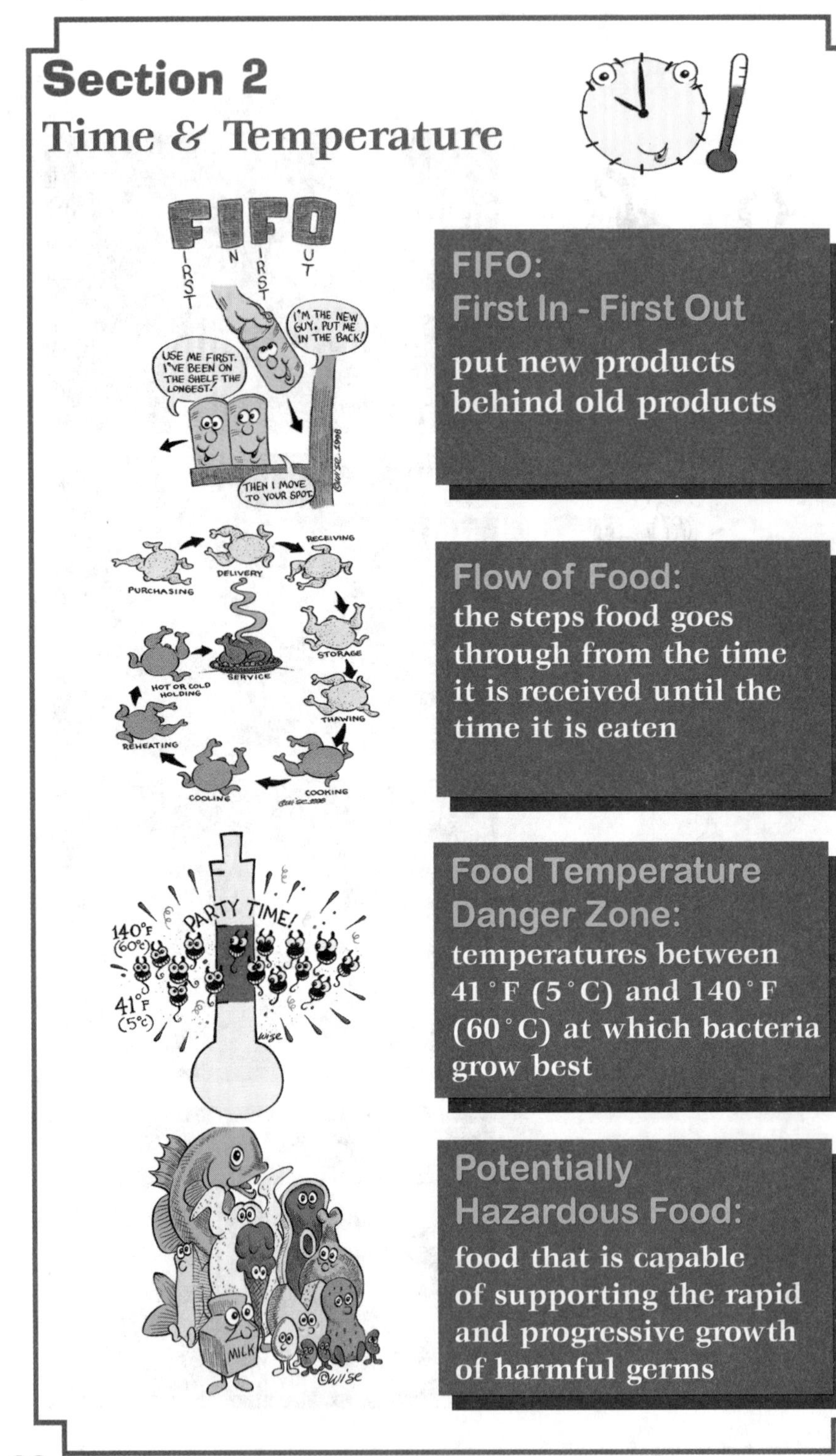

Section 2
Time & Temperature

FIFO: First In - First Out

put new products behind old products

Flow of Food:

the steps food goes through from the time it is received until the time it is eaten

Food Temperature Danger Zone:

temperatures between 41°F (5°C) and 140°F (60°C) at which bacteria grow best

Potentially Hazardous Food:

food that is capable of supporting the rapid and progressive growth of harmful germs

Section 3
Cross Contamination

Cross Contamination: transfer of harmful germs between items

Sanitize:

reducing the number of harmful germs to a safe level

Section 4
Cleaning & Sanitation

Clean:

you cannot see dirt or food pieces

Sanitary:
the number of harmful germs has been reduced to a safe level

Notes:

A note to managers, supervisors, and food workers:

Pictures and drawings provide fast and easy-to-understand information for employees. In this book, the authors have organized concepts for any food worker to find in a hurry including:

- What causes foodborne illness
- How to prevent contamination of food
- The importance of personal health and hygiene
- How to use thermometers when checking the temperature of food
- Methods of thawing, cooking, cooling, and reheating food
- How to prevent cross contamination
- Methods to clean and sanitize equipment and utensils
- And finally, a quick reference chart of times and temperatures required to keep food safe.

John Wise Drawings

John Wise, the artist who drew the pictures, is a master at "boiling" a concept down to just a few lines. Thanks to John, you have in your hands a most entertaining piece of information about a very dry subject. Who can resist the "march of the microbes" or how to "wash those germs right off of your hands"? Use it and enjoy enhancing your knowledge about germs, food, cleaning, and sanitizing.

To contact John, send email to jwise2@tampabay.rr.com